WHAT'S INSIDE ME?

My STOMACH

by Sloane Hughes

BEARPORT
PUBLISHING

Minneapolis, Minnesota

Credits: Cover, all background, © Piotr Urakau/Shutterstock; cover, 7, 23 © BlueRingMedia/Shutterstock; cover, 4, 9, 14, 16, 22 stomach illustration © Shutterstock; 4 Oksana Shufrych/Shutterstock; 5 Dean Drobot/Shutterstock; 6 Nerthuz/Shutterstock; 8 Images By Kenny/Shutterstock; 9 Iconic Bestiary/Shutterstock; 10 Monkey Business Images/Shutterstock; 11, 14 (food) Inspiring/Shutterstock; 11 Volha Kratkouskaya/Shutterstock; 12 NAR studio/Shutterstock; 13 Krakenimages.com/Shutterstock; 15 wavebreakmedia/Shutterstock; 16 (intestine) Grow studio/Shutterstock; 17 Ulia Solovieva/Shutterstock; 18 Tatjana Baibakova/Shutterstock; 19 all_about_people/Shutterstock; 20, 21 Monkey Business Images/Shutterstock.

President: Jen Jenson
Director of Product Development: Spencer Brinker
Senior Editor: Allison Juda
Associate Editor: Charly Haley
Designer: Oscar Norman

Library of Congress Cataloging-in-Publication Data is available at www.loc.gov or upon request from the publisher.

ISBN: 978-1-63691-445-9 (hardcover)
ISBN: 978-1-63691-452-7 (paperback)
ISBN: 978-1-63691-459-6 (ebook)

Copyright © 2022 Bearport Publishing Company. All rights reserved. No part of this publication may be reproduced in whole or in part, stored in any retrieval system, or transmitted in any form or by any means, electronic, mechanical, photocopying, recording, or otherwise, without written permission from the publisher.

For more information, write to Bearport Publishing, 5357 Penn Avenue South, Minneapolis, MN 55419. Printed in the United States of America.

CONTENTS

The Inside Scoop 4
Starring the Stomach 6
The Munching Mouth 8
Getting Started 10
Mighty Muscles 12
Lovely Liquid 14
Helpers of All Sizes 16
A Happy Stomach 18
Keeping Things Moving 20
Your Busy Body 22
Glossary 24
Index 24

THE INSIDE SCOOP

Your body is a super machine that keeps you moving, learning, and having fun. But how does it work? The secret is inside.

When you're hungry, your stomach might start to growl. That's your body telling you it's time to eat! Let's take a closer look.

STARRING THE STOMACH

Your stomach is part of the **digestive system**. It helps your body take care of the food you eat.

The stomach is J-shaped.

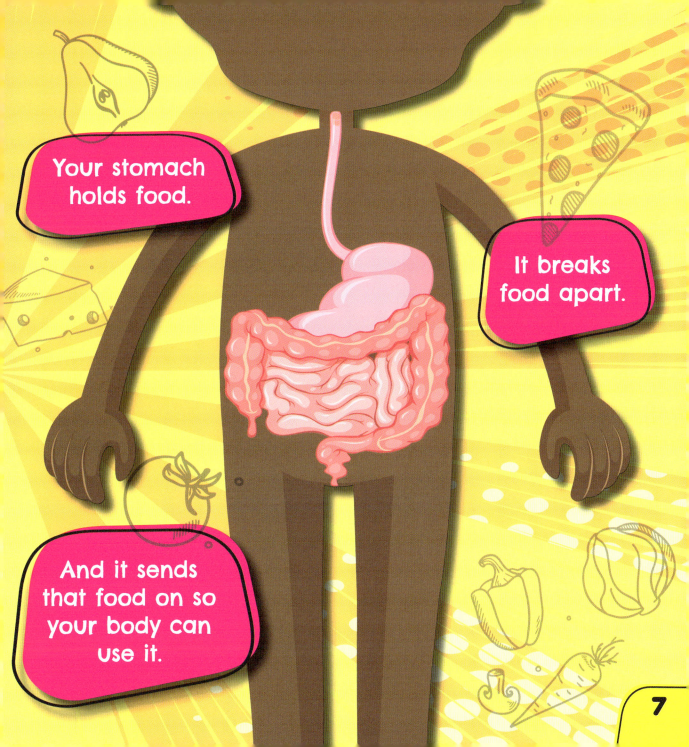

THE MUNCHING MOUTH

The stomach gets a head start when it comes to breaking down food. That job begins in your mouth!

GETTING STARTED

A **valve** at the top of your stomach opens to let food in. Then, it closes to make sure the food stays put.

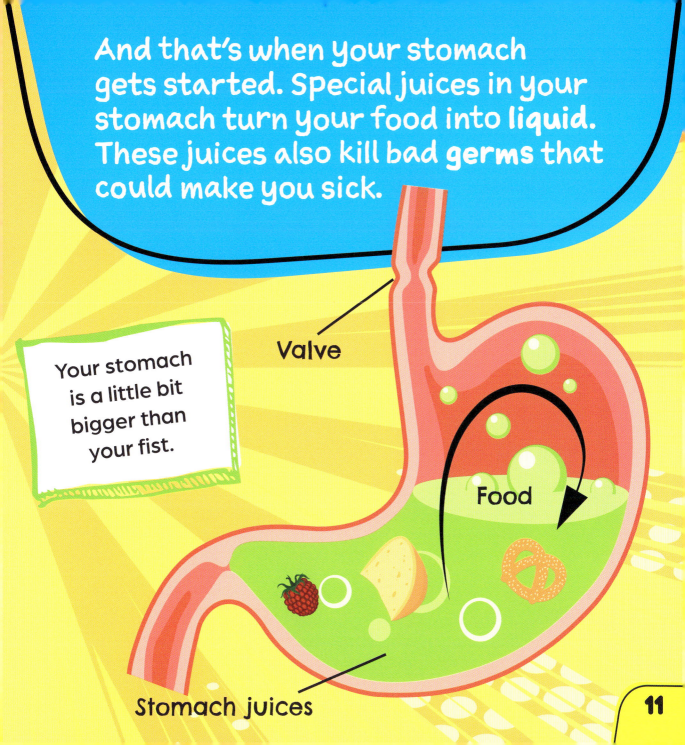

MIGHTY MUSCLES

Muscles go to work inside your stomach, too. They mash food into smaller and smaller pieces.

The stomach is very bendy. It can stretch and pull as it works.

Some foods break down fast. Others are slower. Food usually stays in your stomach between one and five hours.

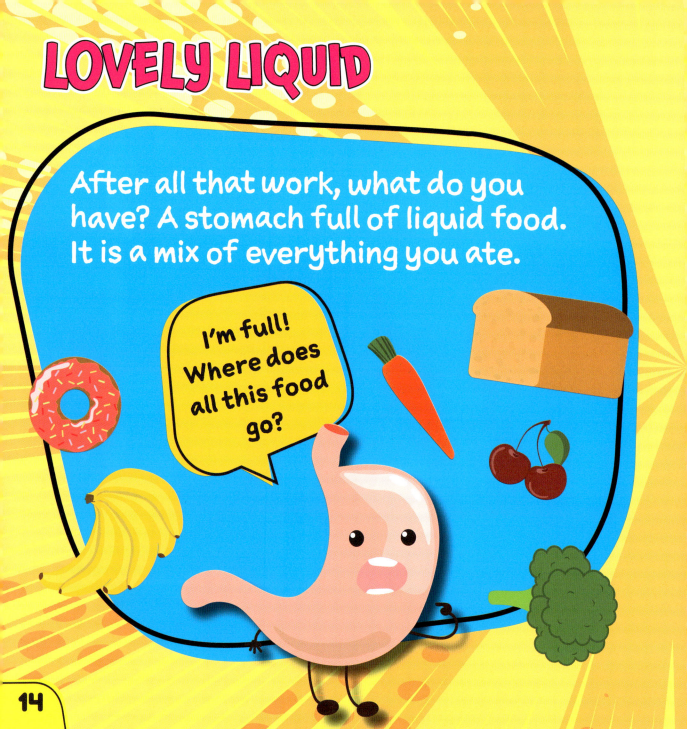

Some of the liquid food has important stuff that helps your body work. You can get **energy** out of it. But your stomach needs a little help for that.

HELPERS OF ALL SIZES

Your stomach has helpers that are large and small. Food goes from your stomach to the small intestine (in-TES-tin) and large intestine. These helpers take **nutrients** and water from your food.

"Thanks for finishing the job, intestines."

"You bet!"

16

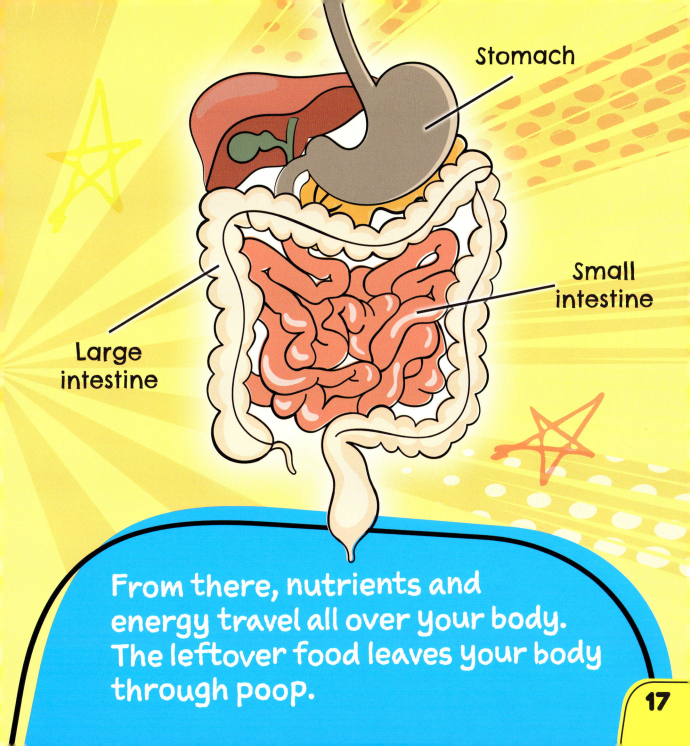

Stomach

Small intestine

Large intestine

From there, nutrients and energy travel all over your body. The leftover food leaves your body through poop.

A Happy Stomach

A happy stomach makes for a happy body! You can help your stomach by eating foods that are easy for it to break down. Whole grains, fruits, and dark green veggies are good for it.

KEEPING THINGS MOVING

Moving your body often helps keep food moving inside you, too. Dance to your favorite song or take some time to stretch. There are plenty of ways to move your body to move your belly!

YOUR BUSY BODY

Your stomach is an important part of the super machine that is your body. It works with lots of other things inside you. Together, they keep you going every day!

You're beautiful on the inside!

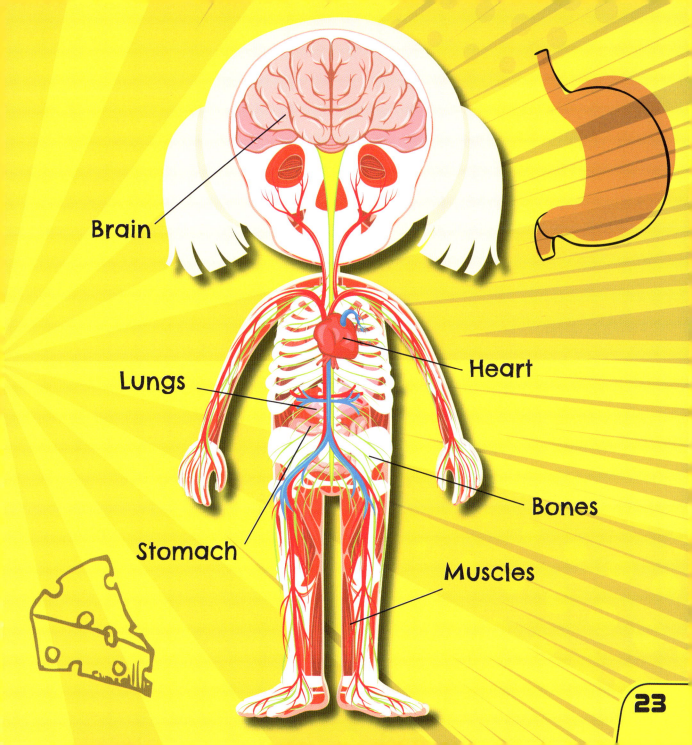

GLOSSARY

bacteria tiny life forms that cannot be seen by the human eye

digestive system the group of organs that help break down food so the body can use it

energy the power needed by all living things to grow and stay alive

germs tiny living things that can make people sick

liquid a thing that flows and has no set shape

muscles parts of the body that help you move

nutrients things in food that are needed to grow and stay healthy

saliva a liquid in the mouth that helps you chew and swallow

valve a part of the body that controls the flow of movement

whole grains the seeds of plants such as wheat that are kept whole and used as food

INDEX

bacteria 19
digestive system 6
energy 15, 17
germs 11, 19

juices 11
large intestine 16–17
liquid 11, 14–15
muscles 12, 23

saliva 9
small intestine 16–17
teeth 9
valve 10–11